Ecosystems & Environment

ANN FULLICK

Heinemann
LIBRARY

First published in Great Britain by Heinemann Library,
Halley Court, Jordan Hill, Oxford OX2 8EJ,
a division of Reed Educational and Professional Publishing Ltd.
Heinemann is a registered trademark of Reed Educational & Professional
Publishing Limited.

OXFORD MELBOURNE AUCKLAND
JOHANNESBURG BLANTYRE GABORONE
IBADAN PORTSMOUTH NH (USA) CHICAGO

8890 $\left(\text{STT}\right)$

Designed by AMR
Illustrations by Art Construction
Printed in Hong Kong

03 02 01 00 99
10 9 8 7 6 5 4 3 2 1

ISBN 0 431 07666 9

British Library Cataloguing in Publication Data
Fullick, Ann
Ecosystems and environment. – (Science topics)
1.Biotic communities – Juvenile literature 2.Environmental
sciences – Juvenile literature
I.Title
577

Acknowledgements
The Publishers would like to thank the following for permission to reproduce photographs:
Ancient Art & Architecture pg 16; Mary Evans Picture Library pg 29; FLPA Images of Nature pgs
4 /K. Aitken, 6 /F. Hoogervorst/Foto Natura, 8, 9 /Fritz Folking; Holt Studios pgs 11 and 24 /Nigel
Cattlin, 18 /Mary Cherry; Morris, Frances & Richard pg 25; NHPA pgs 7 /Hellio & Van Ingen, 12
/David Middleton, 20 /Nigel J. Dennis; PA News Photo Library pg 27 /Roslin Institute; Quadrant
Picture Library pg 23; Science Photo Library pgs 5 /CNRI, 13 /Eye of Science, 17 /Claude Nuridsany
& Marie Perennou; Tony Stone Images pg 22 /Ken Fisher; Topham pg 19 /Associated Press.

Cover photograph reproduced with permission of NHPA/B. Jones and M. Shimlock.

Our thanks to Geoff Pettengell for his comments in the preparation of this book.

Every effort has been made to contact copyright holders of any material reproduced in this
book. Any omissions will be rectified in subsequent printings if notice is given to the Publisher.

For more information about Heinemann Library books, or to order, please phone 01865 888066, or
send a fax to 01865 314091. You can visit our web site at www.heinemann.co.uk

Any words appearing in the text in bold, **like this**, are explained in the Glossary.

Contents

Habitats and ecosystems

Every living thing has a home, and all living things interact with their **environment** (surroundings) and with each other. By finding out about these relationships, we can understand more about the way our planet works.

Home sweet home

Almost everywhere in the world is home to at least one type of living organism. Some of these habitats are large and easy to think of – a coral reef, a rotting log and a meadow by a stream are examples. Other habitats are more unusual and far less obvious. For example, the clown fish lives amongst the poisonous stinging tentacles of large sea anemones, while another fish, *Tilapia grahami*, survives in the warm mineral waters of Lake Magadi in Kenya, at temperatures of up to 43°C.

In an area of the sea like this, there are a number of different **habitats** providing homes for a wide range of living organisms.

A wider view

If we look at a single plant or animal in its habitat, we soon discover that its home is only a small part of a particular area. Within that area there are lots of other species of animals and plants, each living in its own habitat. Very often these animals and plants interact with each other. For example, in a woodland the trees are the habitat of many different species of birds, mammals and insects. Fallen trees are the habitat of other creatures, like woodlice, bacteria and fungi.

The woodland floor is also home to many plants and animals. Some of the woodland animals eat bits of the plants, while others eat animals. The sort of plants that can grow in a woodland depend on the amount of rainfall in the area, the type of rocks and soil, the amount of light and the normal temperature range. Factors like these also affect the types of animals found, both directly and indirectly, as a result of the plants that are able to grow.

This complex interaction of many different species of living organisms with the **abiotic** (non-living) features of their home is known as an **ecosystem**. The study of ecosystems is called **ecology**. Abiotic factors are very important in an ecosystem. They not only determine the kind of organisms living in an area, they also affect the numbers of each type.

The human habitat

A surprising number of organisms prefer to live on human beings rather than on any other mammal. Every healthy human being is home to countless bacteria. They live all over the surface of the skin and in our gut. Mites and other tiny invertebrates swing through our body hair like the inhabitants of a microscopic jungle. We are also hosts to parasitic worms, fungi, lice, fleas and many other organisms that wait for a breakdown in our body's defences. As soon as this happens they can make themselves at home on or in our warm, food-rich bodies.

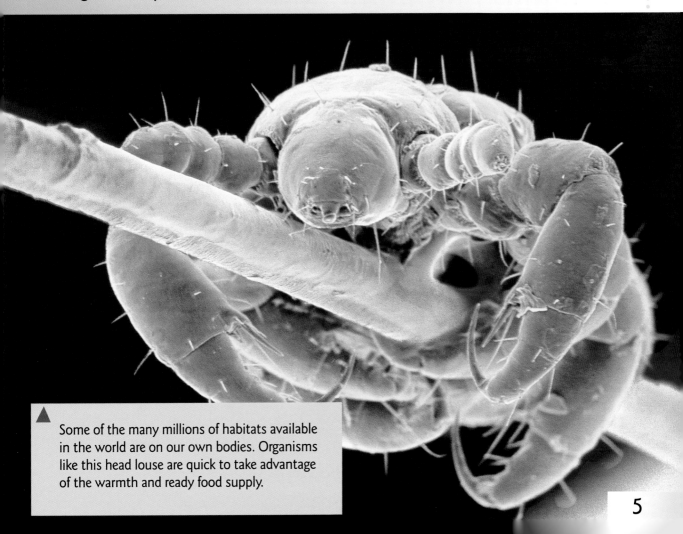

▲ Some of the many millions of habitats available in the world are on our own bodies. Organisms like this head louse are quick to take advantage of the warmth and ready food supply.

Food chains

All living things need food to give them energy. For most living organisms the energy from food originally came from the Sun, either directly or indirectly.

Powerhouse of the planet

Plants are often taken for granted, but they are the true powerhouse of our planet. For living things to live, grow and reproduce, they need energy. Plants and some **bacteria** are the only living organisms that can harness the Sun's energy and use it to provide themselves with food.

They do this using a process called photosynthesis. Energy from the Sun is captured using a special green colouring called **chlorophyll**, which is found in **chloroplasts** in the cells of leaves. Once captured, the energy is used to join carbon dioxide from the air with water from the soil to make sugar. Oxygen is released as a waste product. The sugar is then used to provide energy for the plant and raw materials for building new plant cells. Because of this ability to build new biological material (**biomass**), plants are known as producers.

All of the biomass present in a tree comes from photosynthesis. Food and oxygen are produced in this one amazing process. Plants are often taken for granted, yet they are vital for life on Earth.

Links in a chain

Because of their unique ability to produce new biomass, plants are used as food, either directly or indirectly, by all the different types of animals on Earth. Animals are called **consumers** because they have to eat other organisms to live. They either eat plants themselves or eat animals that are linked back to other animals that eat plants. These feeding relationships, which always lead back to plants, are known as **food chains**.

Herbivores eat only plants and are known as **primary consumers**. Animals that eat herbivores are called **secondary consumers**. Secondary consumers are themselves eaten by tertiary consumers. Animals that hunt and eat other animals are known as **predators**, while the hunted – mainly herbivores and small carnivores – are known as **prey**.

Food chains can be very short. For example grass is eaten by cows and cows are eaten by people. They can

▲ A **carnivore** may not seem to have much to do with plants, but the moorhen doomed to be a meal for this fox was busy feeding on plants only a short time ago. Without those plants there would be no food for the meat-eaters.

also be quite long, moving through many organisms of increasing size. Whatever the length of a food chain it will always start with a green plant. All food chains also include decomposers. These are specialized consumers, often fungi or bacteria, which break down the tissues of a plant or animal when it dies and so return the nutrients to the soil.

Freedom from the Sun

A very tiny group of living organisms do not rely on green plants and the Sun to supply their energy. Specialized **bacteria** known as methanobacteria produce energy from hydrogen and carbon dioxide. Methane gas is also produced, but as a waste product. Other bacteria use sulphur or hydrogen sulphide, ammonium ions and nitrites as

energy sources. The strange biochemistry of these organisms allows them to live in difficult habitats, such as swamps, with extremes of temperature, salt concentration and pressure. However, it has not allowed them to take the place of plants in providing energy for all other living organisms on the planet.

Food webs

A **food chain** gives us a simple model of the feeding relationships between plants and animals. However, in real life the situation is almost always more complicated.

Living webs

Within an **ecosystem** there are many different types of plants and animals living in a particular area. When we construct a food web we identify the feeding relationships between the **producers** and **consumers** in an ecosystem, and show how the animals and plants are interlinked.

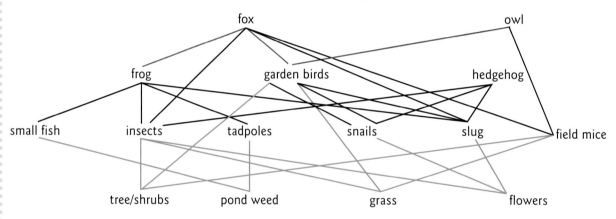

A food web from the garden: almost everything can eat – or be eaten – by at least two or three other organisms.

Within any ecosystem, animals and plants are linked together in a complex web of life. Even though an ecosystem like this garden is influenced by human activity, it still has a complex food web.

Natural buffering

If each animal had only a single food source then the whole balance of nature would be very delicate indeed. For example, if a **herbivore** ate only one type of plant or a **carnivore** hunted only one type of **prey**, they would find it difficult to survive.

However, as food webs show, this is rarely the case. In theory, if all the slugs in our garden were removed by a keen gardener, then more plants would survive. In fact, there would just be more food left for the greenflies, ants and other organisms that eat plants. The many garden birds, hedgehogs and other animals that eat slugs would not starve to death. Instead of the slugs they would eat more of the other species they like. As a result, there would probably be fewer worms, beetles and other creepy-crawlies in the garden.

An environmental event that affects one organism does not have as large an effect on a whole ecosystem as we might expect. This is because of the buffering effect of the many organisms that are all interacting.

Panda problems

Pandas are a well-known symbol of conservation, not least because they are themselves dangerously close to **extinction**. Part of the problem for pandas is that they are not really part of a food web. Instead, they are part of a simple food chain in which pandas eat bamboo and little else. When the bamboo plants all flower and die, which happens every hundred years or so, and if the pandas' habitat is destroyed, they immediately face starvation. On looking at how vulnerable the panda is, we can certainly appreciate the buffering effect of food webs. Fortunately, they support the majority of living organisms.

Pandas are beautiful but vulnerable. These rare animals have the guts and teeth of a carnivore but eat a diet of almost entirely one plant. No one is sure why pandas eat only bamboo, but it certainly makes it much harder for them to survive.

Pyramids of life

A close look at the organisms making up a **food chain** or **food web** reveals an important principle of **ecology** – in almost every community there are more plants than **herbivores** and more herbivores than **carnivores**.

Pyramids of numbers

Animals and plants that live in any particular **ecosystem** are often linked by food chains or food webs.
For example:

```
grass → rabbit → fox
sea plants → fish → seals → polar bears
```

Although food chains like these show us the feeding relationships between the organisms, they do not give us any idea of how many of the different organisms there are. If we count or quantify the number of organisms in a food chain, or in the chains that make up a complex web, we can build up a pyramid of numbers.

The numbers in many food chains confirm what we expect: many plants at the bottom of the pyramid support fewer plant-eaters, which in turn feed a smaller number of larger **secondary consumers**. These may in turn feed an even smaller number of top **predators**.

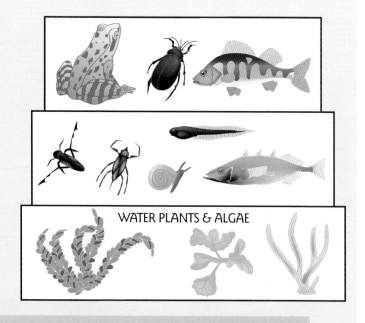

▲ Whether we look at a single food chain or a whole food web, the numbers of living organisms supported at each level of a pyramid of numbers tends to get smaller.

Biomass is better

Sometimes there is a problem when we use pyramids of numbers to help analyse an ecosystem – the numbers can be greater at the top, so there is no pyramid.

▶ Thousands of aphids feed on just one rosebush. The aphids themselves act as food for ladybirds and other predators. The pyramid of numbers for a food chain like this is a very strange shape indeed!

However, when we look at the amount of actual biological material (**biomass**) at each level of a food chain, we always get the picture we expect. We see the amount of biomass falling as we move from plants to **herbivores** to **carnivores**. Why do the levels of biomass fall in this way? Where does all the 'lost' material go?

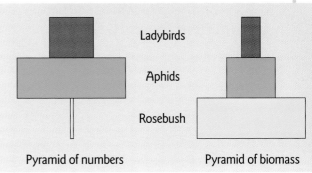

Ladybirds

Aphids

Rosebush

Pyramid of numbers   Pyramid of biomass

Much of the sun's energy captured by plants in **photosynthesis** does not get through the food chain. A great deal of the plant material eaten by herbivores is passed out of their bodies as **faeces** because they cannot digest **cellulose**. Carnivores also leave parts of their **prey** uneaten and get rid of undigested material as faeces. The food that animals do digest is used to provide them with energy for movement and, if the animal is a mammal, to produce body heat. Only a small amount of the food energy is turned into new animal, to be passed on to the next animal in the chain.

Measuring biomass

Counting numbers of living organisms in a food chain can be difficult, but measuring biomass is even harder. If the animals and plants are alive their biomass still contains water. A measure of their wet biomass can be very inaccurate. For example, it will be affected by how much water the animals have drunk. Measuring dry biomass gives an accurate picture of a food chain. Unfortunately, to get the biomass dry, the organisms have to be killed and dried. This destroys the chain being studied!

Making the best of it

On any small area of the Earth's surface we can find a whole range of different **habitats**. Wherever life is possible, it will be found. Animals and plants have **evolved** over millions of years to take advantage of any opportunities that come their way.

Survival!

Animals and plants need to be able to survive and reproduce in the **environment** in which they live.

Many have developed adaptations that enable them to take advantage of the food sources available.

Desert organisms have many different ways of surviving. Cacti have long, deep roots, and stems full of cells that can store water. The leaves of cacti are thin spines, which reduces the water loss. A special form of **photosynthesis** means that they only need to take in carbon dioxide during the night, and so keep their pores closed during the day to prevent water loss. All these adaptations mean that cacti do well in conditions where most plants would wilt and die.

Other desert plants survive by avoiding most of the dry conditions, living as **dormant** seeds that only start to grow after rain has fallen. Once wet, they flower and set new seeds within a few days. The plants will survive even if there is no more rain for several years.

The hot, dry conditions of a desert environment are too challenging for most living things. The organisms that can survive and reproduce in these extreme conditions have special adaptations that make their survival possible.

In the animal world, camels are well known for their humps of fatty tissue. These provide food for the camels for many days. Camels are able to manage for long periods without water. They have a special system for cooling the blood as it flows to the brain and **enzymes** that tolerate far higher temperatures than those of most mammals. Their thick insulating coats trap a layer of air to keep out the heat of the day and the cold at night. Their ears can close to prevent sand getting in and their special wide feet are excellent for walking on sand. Camels are perfectly adapted to their environment. Other desert animals have different adaptations. Many of them simply avoid the heat of the day, living in underground burrows and only emerging during the evening, night and early morning. Most desert animals are adapted to survive with little water.

Habitats everywhere

Living things have made their homes in many different environments – from the cold regions of the Arctic and Antarctic to the arid deserts of the world, from the heights of the Rocky Mountains to the depths of the Marianas Trench (the deepest known part of the ocean). Some of their adaptations are fairly common – such as fur, giving them an insulating layer, or different shaped mouths, beaks and teeth that allow them to catch and eat different types of food. Other adaptations are quite bizarre.

The mole-rat of East Africa is an unusual creature. It spends its entire life underground and only eats the roots of grass. Family units live together in a maze of underground tunnels with special chambers for different activities. Their adaptations to this life are dramatic. They are blind, naked, sausage-shaped animals with grey, wrinkled skin. They have huge incisor teeth with which they can move soil while their mouths are shut. By eating roots and living completely underground they avoid dangerous **predators**, unlike the prairie dogs and rabbits that feed on grass at the surface.

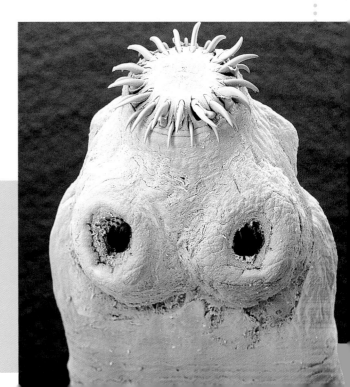

▶ The ultimate adaptation? Parasitic organisms like this tapeworm live inside the body of another organism. Warm and protected from predators, they feed on already digested food. To avoid being digested themselves, they have to be adapted for hanging tightly onto their host. These parasites have certainly found a very exclusive habitat.

Keeping up with change

Organisms must be adapted for the **habitat** in which they live. Sometimes this involves adapting to cope with change. Changes in the environment can test an organism to the limits. Those which cannot cope often die as a result. The adaptable ones survive and breed.

All change!

Some organisms live in very stable **environments**, but others have to cope with constant change. One of the best examples of organisms coping with change is in the tidal zones of a coastline. Within 24 hours the tide comes in and goes out at least twice. As a result, the plants and animals are covered by moving water for part of the time. Next, they are exposed either on the sand or in the shallow water of a rock pool. As the water in a rock pool evaporates in the heat of the sun, its salinity (concentration of salt) becomes far greater than that of normal sea water.

▼
Habitats such as rock pools are a real challenge to the organisms that live in them. A rock pool forms a **microhabitat**. Depending on its position on the beach, it can be anything from a few minutes to a few hours out of the moving sea every day. The trapped plants and animals have to cope with the changes in temperature and salinity.

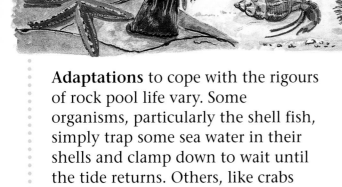

Adaptations to cope with the rigours of rock pool life vary. Some organisms, particularly the shell fish, simply trap some sea water in their shells and clamp down to wait until the tide returns. Others, like crabs and seaweeds, have **evolved** tissues that can cope with the increase in temperature and salinity that occurs when a rock pool is exposed to the sun for some time. Any organisms trapped in a rock pool that are not adapted to cope are likely to die before the tide returns.

The changing seasons

Near the equator the weather stays very similar all through the year. But in many other parts of the world, living organisms have to cope with dramatic changes in conditions. Within one year in some states in the USA, plants and animals have to adapt to changes of around twelve hours in the amount of daylight and to temperatures that range from –10°C in winter to up to 30°C in summer. The adaptations that enable organisms to cope range from physical characteristics to big changes in behaviour.

Dormancy: many plants lose their leaves and shut down for the winter. Others exist as a dormant seed, bulb or tuber underground.

Hibernation: some animals build up large fat stores ready for the cold weather. During the winter they drop their body metabolism to a very low level and sleep. Others indulge in partial hibernation, waking every so often to feed on stores of food.

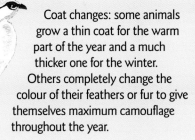

Coat changes: some animals grow a thin coat for the warm part of the year and a much thicker one for the winter. Others completely change the colour of their feathers or fur to give themselves maximum camouflage throughout the year.

Migration: many birds and some mammals travel vast distances to avoid adverse conditions, and migrate back again when the seasons change.

Biological clocks

Almost all living organisms have to cope with the 24-hour rhythm of day and night. Many organisms have their own internal rhythms which are linked to this 24-hour cycle. These are known as biological clocks, and they almost all keep a **circadian rhythm**. Our human clocks tell us when to sleep. They also control the levels of many chemicals in our bodies.

Agents for change

Most of the **adaptations** we see in animals and plants have **evolved** in response to changes in the natural world around them. Changes in the **environment** are not always the result of natural phenomena. There is a new agent for change – the human race.

SCIENCE ESSENTIALS
People often have an effect on the environment.
Animals and plants may be affected by the actions of people and may need to adapt to cope with the changes.

The human factor

Throughout the long history of life on Earth, species of animals and plants have been in a constant process of adapting and changing to survive. Species which are not well adapted become extinct. When a new **habitat** becomes available, animals and plants will adapt to fill it.

Usually, environmental changes occur slowly, over a long period of time. This gives organisms a chance to adapt over many generations.

In the relatively short time since human beings have evolved, we have had a major impact on the environment and the animals and plants that share our planet. We have reached a point where we can bring about major environmental changes rapidly. This means that other organisms need to adapt more quickly than ever before if they are to avoid **extinction**.

▲ When people began to hunt in groups and use weapons they became very effective killers. It is possible that the efficient hunting of early people helped wipe out species, such as the woolly mammoth and sabre tooth tiger, which could not adapt to cope with the attack.

Biston betularia and the Industrial Revolution

The peppered moth, *Biston betularia*, is found naturally in two different colour forms, a light speckled colour or a black form. Until the 19th century the pale form was by far the most common in Britain. On tree bark it was almost invisible to the birds that wanted to eat it, and so was much more successful than its darker relative. However, after the Industrial Revolution, smoke pollution from factories darkened the trees, particularly in the north of Britain, and the number of black moths increased rapidly. On the dark tree trunks it was the turn of these dark moths to be better camouflaged from the predatory birds. In more recent years pollution has been better controlled and the tree trunks are pale again. As a result, there are once again more light-coloured peppered moths than dark ones to be found.

▶ On a clean tree trunk we can see which of the two forms of peppered moth is better camouflaged. A change in the environment due to human activity also resulted in a change in the most common form of peppered moth.

Helping out the frogs

We hear a lot about human destruction of habitats and its effect on different types of animals and plants, but sometimes people do just the opposite and help a species to survive. In the tropical rainforests of Puerto Rico, for example, there is a species of frog that feeds on insects at night, hiding from predators during the day. Their food is plentiful so scientists were puzzled when they found that the numbers of frogs were low. Small bamboo shelters were built so that the frogs had more hiding places. As a result, the frogs were able to breed more successfully and their numbers rapidly increased. All over the world bird boxes and bat boxes are used to do the same thing – increase the breeding success of different animal species.

Good or bad?

We are constantly affecting the world around us. Is our influence good or bad? It is easy to show how people are damaging the environment, but there are many examples of humans in positive roles as well.

The human managers – good...

Sand dunes are vulnerable **environments** in coastal regions around the world. All too easily they are damaged and blow away. Conservation work using fencing and planting marram grass can stabilize the dunes and prevent their destruction. Many other positive actions are taken by people all over the world, too. These include setting up national parks and conservation areas, providing nesting boxes and managing woodlands so that new trees grow before the old ones die.

All over the world there are ecosystems and species of animals and plants which owe their continued existence to the support of human communities. At this tree nursery in Ghana, local workers are looking after young seedlings as part of a forestry project.

...or bad?

The production of food is vital. However, the pesticides and herbicides used to help grow crops can be devastating to wildlife if they build up in food chains. Also, the fertilizers used to increase crop yields can cause problems in rivers, ponds and lakes, sometimes killing the animals that live there.

Many developing countries, such as Brazil, are cutting down rainforests to provide wood for developed countries, such as the USA and UK, to make furniture, ornaments and even toilet seats. The cleared land is then often used to farm cattle to provide beefburgers for the **developed world**.

Most people in these developing countries are desperately poor. They are exploiting the only resource they have to try and get food, medicine and education for their families. Yet the effect of rainforest destruction on the whole planet is potentially devastating. This, together with the problems of **acid rain** and the build-up of greenhouse gases from our car and factory fumes (the **greenhouse effect**), show how serious the effects of some human activity can be on the Earth's ecosystems.

A Unknown to anyone, waste from a plastics factory in Minamata, Japan, contained high levels of mercury. It was pumped out into the bay and built up in the mud at the bottom of the sea.

B In the bay, levels of mercury built up in shellfish and fish, but did not kill them.

C Local fisherman fed both their families and their cats on the fish and shellfish they caught. The cats became ill first, gradually becoming paralysed and finally dying. No one could understand why.

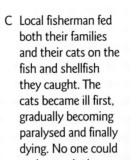

D Finally, people showed the same symptoms. More than 60 people were permanently disabled, 44 people died, and babies were born deformed before the source of the mercury poisoning at the beginning of the food chain was found and stopped.

▲ The Minamata tragedy. In the 1950s the effects of human pollution travelled along a food chain and devastated the lives of people in the area.

Biological pest control

Chemical pesticides have caused a lot of **environmental** damage in the last hundred years. On the other hand, without the pesticides many crops would have failed and millions more people would have died of starvation. Increasingly, people are looking for ways to control pests without using toxic chemicals. Biological pest control seems to offer one of the best ways forward.

This involves using an animal (often an insect) or a plant to control pest organisms. For example, in your garden you can find ladybirds and put them on a plant infested with greenfly which the ladybirds will eat. On an international level, research is being done to find suitable, safe organisms to control pests that damage crops or destroy natural environments on a large scale.

Populations – plants and animals

Ecology mainly looks at the factors that affect the distribution and abundance of living organisms. This involves studying the success or failure of **populations** rather than single animals or plants – large numbers give us a much more accurate picture of the way things work.

What is a population?

A population is a group of similar organisms living in a particular habitat at the same time, producing and consuming food, using up nest sites and generally making use of the surroundings. In theory any population will double in a given period of time. If conditions are ideal it will continue doubling in size to show **exponential growth**. Fortunately, for the balance of life on Earth, ideal conditions rarely exist, and so populations grow at fairly slow rates!

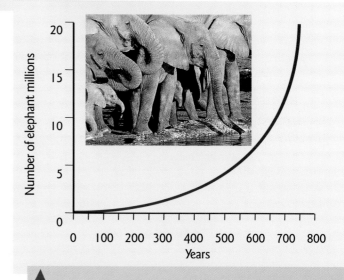

Charles **Darwin** calculated that a single pair of elephants breeding in ideal conditions would produce 19 million descendants in just 750 years.

Ups and downs of life

The number of birds in a garden can vary greatly from one year to another. The number of coyotes in one area of wilderness can be very different from that in another area only a few kilometres away. Basic factors affect population numbers, such as the type of rocks and soil, the weather, the temperature, and the amount of water and light available. The weather has a general effect. For example, populations of cacti are generally found where the weather is hot and dry, but snowdrops grow in **temperate climates**. Weather can also have very local and severe effects. For example, all over the world hurricanes and tornadoes not only devastate human populations, they also bring death and destruction to tree populations, along with the animals living in and on them. But in the years that follow, the numbers of fungi, wood-eating insects and light-loving plants increase sharply, filling the gaps left by the trees.

Competition has a big effect on population numbers. It may occur between different members of the same species, or members of different species. The interactions between predators and prey also affect population size. In theory this is simple. The prey population feeds and grows, providing lots of food for the predators who in turn feed, reproduce and increase in number. The increased number of predators eat more prey, so the prey numbers fall. Shortly afterwards predators begin to starve to death and the predator numbers fall, allowing prey numbers to rise again... Cycles of numbers like this can be seen in the real world. However, a situation is not always as simple as it seems.

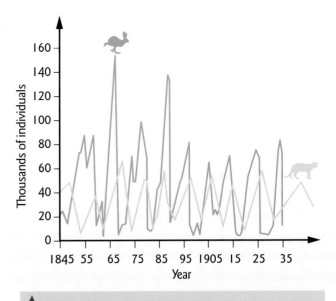

At first it appears as if the population 'peak and crash' cycles of the snowshoe hare and lynx are closely linked together in a classic predator-prey cycle. In fact, the snowshoe hare population rises and falls in the same way in areas where there are no lynxes! The changes are due to fluctuations in the weather and the numbers of plants that the hares need for food.

Population patterns

Some populations of animals and plants have a very high density (lots of the organisms live very close to each other).

Others have a very low density (only a few organisms live in a big area).

Uniform: each animal defends its own territory.

Random dispersal: groups of individuals are scattered randomly when resources are plentiful and there is no sign of antagonism.

Clumped: where a herd of animals or groups of plants are found in gatherings around resources such as food or water. This is the most common arrangement.

Depending on both the population density and type of organism, populations of animals and plants are usually spread out in one of these three patterns.

The human population

Throughout the history of life on Earth **populations** have grown and declined. However, for the first time we now have a population that simply keeps growing.

A human problem

Human beings have lived on Earth for only a short period, compared to the history of our planet. For thousands of years the human population grew very slowly. Later, improvements in weapons for hunting, tool-making skills and crop-growing led to increases in the population. This growth was largely cancelled out by deaths due to catastrophes, diseases and wars. At the time of the agricultural revolution in the Middle Ages, it is estimated that there were only about 133 million people in the world. In recent centuries our ability to grow food and treat diseases has developed rapidly. As a result, the human population has grown enormously. It now has to be measured in billions.

▼ If any other species on Earth experienced a population explosion of the size seen in humans, scientists would predict that an equally dramatic crash would follow. Is that our future too?

A force to be reckoned with

The vast human population has spread across the surface of the Earth to take advantage of almost every land **habitat**. To maintain our ever increasing population we will need to use all of the Earth's resources very carefully.

The biggest problem we face is in the production of food. It is not simply a case of growing enough – we do that already. Difficulties lie in producing food where it is needed and in distributing it around the globe. Many people in the **developing world** are largely vegetarian. If the **developed world** ate less meat there would be a lot more food to go round as the amount of plant food needed to fatten one animal for meat would keep quite a few people from starving.

Another problem is the growth of enormous cities, particularly in the developing world. It is estimated that soon 50 per cent of the world's population will live in cities. Overpopulated cities cause hygiene and sewage management problems, poor health and disease. As people take up more and more space, habitats will be lost and other animal species pushed out.

Pollution of land, air and water is bound to increase as the human population grows. Our only real hope is that population growth can be slowed down.

▶ More people means more energy will be used. **Fossil fuels**, such as the petrol we use to run our cars, will not last forever. Also, as these fuels are burnt more greenhouse gases are produced, adding to the **greenhouse effect**.

Bringing down the birth rate

In many developed countries the birth rate has fallen dramatically over recent years. Contraception has become easily available and women have become educated and free to lead independent lives. Food and medicine have improved so relatively few children die. As a result many women have chosen to have fewer children. In many countries the birth rate has dropped below two children per couple, so the population is falling. However, each child in the developed world uses far more resources than one born in a developing country.

Often in many developing countries girls are not educated. There may be ignorance about birth control, or it may be forbidden by religious groups. Also, as many children still die from disease and malnutrition, large families are seen as a kind of insurance. And, in many societies, a large family is a status symbol.

It will take many years before people in the developing world feel able to change their views. Meanwhile the human population climbs ever higher.

Introducing variety

Millions of species of animals, plants and micro-organisms are alive today. Although they are largely built up of similar cells, they demonstrate the wide variety of organisms on Earth.

Natural selection

People look very different from each other, but they also look different from any other species – so do the members of any other species. The differences between members of the same species are known as **variation**.

When conditions are good, most plant and animal members of a **habitat** will survive. If conditions get tougher, however, it becomes a fight for survival. The most likely to survive have, by chance, a feature that helps them overcome the difficult conditions. Having survived the difficult times, they are then able to reproduce successfully when times improve. Their useful survival trait is passed on to their offspring. This **natural selection** of organisms with the most useful survival traits was called 'survival of the **fittest**' by Charles **Darwin**.

▶ Artificial selection has given us cereal crops from flowering grasses, and beef and milk cattle from the original wild steers.

All of the vegetables in this picture have been selected for particular features to fulfil a specific need. Internally they are very similar, but they all look very different.

Artificial selection

Not all variety comes about by random natural selection. For thousands of years people have used selective breeding to develop animals and plants with the features they want. We have made animals tamer, fiercer, smaller and larger to suit our needs. This involves choosing the parents carefully for the desired features, and then selecting the best of the offspring for further breeding until those features are secured.

Nature versus nurture

Variation between organisms is caused partly by the information they inherit from their parents and partly by their environment. It is common sense that the surroundings of a living organism will influence the way it turns out, and scientific evidence backs this up.

Most living organisms vary because they have different genetic material. Sometimes clones appear – individuals which are the result of natural or manipulated **asexual reproduction**. Clones are genetically identical. This should mean that any differences between them are the result of differences in their environment. Plants will grow differently if they are deprived of light or water, and animals, including people, will be affected by their diet and other aspects of their environment.

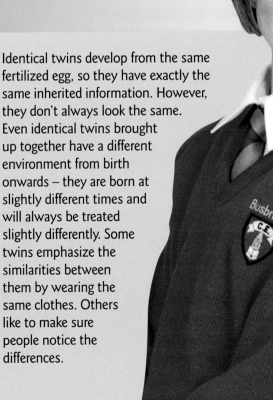

▶ Identical twins develop from the same fertilized egg, so they have exactly the same inherited information. However, they don't always look the same. Even identical twins brought up together have a different environment from birth onwards – they are born at slightly different times and will always be treated slightly differently. Some twins emphasize the similarities between them by wearing the same clothes. Others like to make sure people notice the differences.

'Pharming' in the future

Until recently, farming has involved growing plants and animals to provide for human needs: food, drinks and clothing. But techniques such as **genetic engineering** and **cloning** are changing the way we think about farming.

SCIENCE ESSENTIALS

Genetic engineering gives us the power to change organisms in new and dramatic ways. **DNA** from one organism can be transferred to another, enabling it to make protein from a different species.

Genetic engineering

By changing the **genetic** information in an organism, scientists can affect the way the cells and organs work. This has great potential for farmers all over the world. Putting an extra growth gene into animals and plants can make them grow faster, allowing food to be produced more cheaply. Genes have been put into tomatoes to prevent them going 'squashy' on the supermarket shelves. Genes have been engineered into a variety of crop plants enabling them to produce a natural pesticide. This means that chemical pesticides and their related pollution risks can be avoided. Genetically modified cotton that is exactly the right colour for blue denim jeans is being grown. This saves the company the expense of dying the material and avoids the pollution caused by the dying process.

Along with research into new genetically modified products, scientists are also working to try and make sure they are safe, both for consumers and the **environment**.

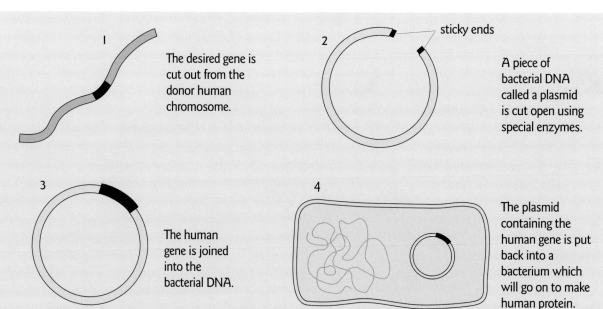

1 The desired gene is cut out from the donor human chromosome.

2 sticky ends

A piece of bacterial DNA called a plasmid is cut open using special enzymes.

3 The human gene is joined into the bacterial DNA.

4 The plasmid containing the human gene is put back into a bacterium which will go on to make human protein.

▲ Genetic engineering involves taking a small fragment of DNA (the genetic material) from one organism and 'stitching' it into the DNA of another organism.

The 'pharm' of the future

Instead of simply farming animals and plants for food and clothing, we are now beginning to 'pharm' them – grow them for the pharmacological substances (drugs) which they can produce after genetic engineering. After the addition of a human gene, **bacteria** that produce human insulin for use by diabetics can be grown. Human genes have also been engineered into mammals such as sheep and cattle. The animals then secrete substances in their milk, such as human growth hormone or blood-clotting factors, for treating serious diseases.

Cloning and beyond

Cloning plants by taking cuttings is a long-established practice, but new cloning skills are now being used in several ways. A plant can be broken down into tiny cell fragments and each fragment grown, making thousands of identical clones. Similar techniques are being researched in animals. If animals that have been engineered to make human proteins can be cloned in this way, the benefits spread quickly.

▶ In 1997 the first cloning of a large mammal, Dolly the sheep, was announced. Dolly has gone on to show how normal she is by producing a lamb of her own.

How far, how fast?

Some people accept these new techniques readily because of the many benefits they bring. Others are more cautious. Most genetic engineering involves marker genes that allow scientists to see if the new genes are in place. These markers are often genes for antibiotic resistance – which means an organism given them will not be affected by antibiotics. Nobody is completely sure that this resistance could not be passed into bacteria, giving rise to antibiotic resistant superbugs capable of causing untreatable diseases. If plants contain pesticide all the time, pests might build up resistance to the poison. There may be problems because people are having to eat the pesticides along with the plants, too. Companies are **patenting** genes and organisms – is it right for a form of living organism to belong exclusively to a company? Society needs to consider these and other questions sooner rather than later, to make sure that the new technologies are a safe way forward.

Extinction

Of the estimated 4 billion species that have lived on Earth during its history, only about 2 million are alive today.

SCIENCE ESSENTIALS

Extinction means the permanent loss of a species of living organism from the surface of the Earth.

Extinction is for ever

The great red elk, the Tasmanian tiger, *Tyrannosaurus rex* – today, all these animals are extinct, even though their names may be familiar.

A landscape like this comes from the mind of the artist. No human has ever seen these dinosaurs alive – dinosaurs disappeared from the Earth long before our early ancestors were even walking upright.

Why do species of animals or plants become extinct? It often seems to be the result of a change in the environment. Many species cope with the change, but others are not so well adapted. Less able competitors do not get as much food or other resources. As a result they are less likely to reproduce successfully. As their numbers fall they will be increasingly pushed out by other, better adapted species. Eventually the species will become extinct in that area. If the environmental change occurs over the whole range of that species, then the whole species may well become extinct.

Mass extinctions

Alongside the normal background rate of extinctions, **fossils** suggest that there have been five **mass extinctions** in the history of the Earth. In one of these, 96 per cent of all the marine **invertebrates** died out. In another, up to 75 per cent of all known plant and animal marine species died out. In the best-known mass extinction, the mighty dinosaurs fell from being dominant animals to being extinct in around 8.5 million years. The effects of these mass extinctions on the evolution of different species should not be underestimated.

After a mass extinction there are large numbers of empty **habitats**, and new animal and plant species rapidly evolve to fill the empty spaces.

The human effect

If a new type of plant or animal evolves it can make other organisms extinct by competing more successfully for the same resources. Human beings have driven more species to extinction than any other. Since people first evolved into social groups, hunting together and using tools, we have wiped many other species off the face of the earth.

Chopping down rainforests and draining marshes and bogs have caused thousands of species to become extinct all over the world. Polluting river and ocean waters, as well as the air around us, has driven many more species to extinction. People are having an enormous effect on the whole environment and the prospects for the future can seem gloomy. However, as we become more aware of the problems we are causing, we can look for ways to reduce or prevent the damage. With people all over the world working together, the chances of saving the Earth for future generations are greatly improved.

It is thought that early people caused the extinction of many of the species they hunted, including the woolly mammoth. Modern people have continued in the same way. The dodo is one of the best-known species to be driven to exctinction by humans. When sailors first landed on the island of Mauritius the flightless dodo was unafraid. Unable to escape, it was hunted to extinction for its meat. Hunting, habitat destruction and pollution continue to cause extinctions around the world every day.

Glossary

abiotic   the non-living features of an **environment**

acid rain   rain that is acidic due to dissolved gases, such as sulphur dioxide, produced by the burning of fossil fuels

adaptations   the special features of an organism that enable it to survive in a particular **habitat**. They may involve the whole organism, parts of the organism or its behaviour.

bacteria   minute single-celled life forms which do not appear to have a proper nucleus

biomass   the amount of biological material

carnivores   animals that eat animals

cellulose   a complex carbohydrate found in plant cell walls

chlorophyll   the green colouring used by plants to capture the Sun's energy

chloroplasts the organelles (tiny compartments) in plant cells that contain chlorophyll

circadian rhythm   the pattern of an organism's life processes which roughly follow a 24-hour cycle

cloning   producing identical offspring from the tissue of one organism

competition   when organisms compete with each other for the same resources

consumers   living organisms that need to eat other organisms to get energy

Darwin, Charles   famous English biologist who lived in the 19th century. He developed and published the theory of evolution based on the survival of the fittest.

decomposers   specialized **consumers**, often fungi or bacteria, that break down the tissues of a dead plant or animal, and so return the nutrients to the soil

developed world   those countries such as the USA, Australia, the UK and much of Europe, which have exploited their resources in order to gain a relatively high standard of living for their populations

developing world   those countries, especially in much of Africa and South America, which are currently preparing their resources for greater exploitation in order to improve their populations' standard of living

DNA (deoxyribose nucleic acid) the chemical that carries genetic information

dormant   when a plant or seed slows its **metabolism** right down to survive difficult conditions

ecology   the study of **ecosystems**

ecosystem   all the animals and plants living in an area, along with things that affect them, such as the soil and weather; also the interaction between many different types of living organisms and the non-living features of their home

environment   an organism's home and its surroundings

environmental   linked to the surroundings

enzymes   protein molecules in living things that speed up or slow down reactions in natural chemical processes

evolved   changed slowly over long periods of time as **adaptations** are passed on by the **fittest** to new generations through breeding

exponential growth   when something rapidly increases in size by doubling in a given time

extinction   the permanent loss of a species of living organism from the Earth

faeces   solid waste passed out of the body

fittest   the most well-adapted organism

food chains   the links between different animals that feed on each other and on plants

food web   model of a **habitat** showing how the animals and plants are interconnected through their feeding habits

fossil   remains of a once-living organism preserved in rock, amber or some other medium

fossil fuels   fuels that have formed over millions of years from the remains of living things

genetic   involving the genes (units of inheritance) in the nucleus of a cell

genetic engineering   changing the **genetic** material of an organism

greenhouse effect   the warming of the Earth's surface as a result of a build-up of greenhouse gases, such as carbon dioxide

habitat   place where an animal or plant lives – its home

herbivores   animals that eat plants

hibernate   when animals build up body fat during the summer and then go into a very deep sleep during the winter, keeping their body temperature very low

inherited   passed on from parents in their **genetic** material

invertebrates   animals without backbones

mass extinctions   periods of time when the great majority of living species become extinct

metabolism   the building-up and breaking-down reactions in the biochemistry of a cell or living organism

microhabitat   a very small, localized **habitat** within a much larger habitat

natural selection   the survival of the **fittest** organisms, and the passing on of their genes through reproduction

patenting   registering an idea or product as belonging to a particular individual or organization

photosynthesis   the process by which green plants make food from carbon dioxide and water using the Sun's energy

population   a group of organisms of the same species living in the same **habitat** at the same time

predator   animal which preys on other animals to obtain food

prey   animal which is preyed on and eaten by **predators**

primary consumers   animals that only eat plants

producers   organisms (plants) that produce new food by **photosynthesis**

pyramids of biomass   a graphic showing the amount of biological material in organisms at each level of a **food chain**

pyramids of numbers   a graphic showing the number of organisms at each level of a **food chain**

secondary consumers   animals that eat other animals

variations   differences between members of the same species

Index